Common Sense Guide to Health and Safety at Work

Subash Ludhra

LONDON AND NEW YORK

First edition published 2014
by Routledge
2 Park Square, Milton Park, Abingdon, Oxon, OX14 4RN

and by Routledge
711 Third Avenue, New York, NY 10017

Routledge is an imprint of the Taylor & Francis Group, an informa business

British Library Cataloguing in Publication Data
A catalogue record for this book is available from the British Library

Library of Congress Cataloging-in-Publication Data
Ludhra, Subash.
Common sense guide to health and safety at work / Subash Ludhra. —First edition.
pages cm. —(Common sense guides to health and safety)
Includes bibliographical references and index.
1. Industrial hygiene. 2. Industrial safety—Law and legislation.
I. Title.
RC967.L86 2014
363.11—dc23 2013037574

ISBN13: 978-0-415-83544-2 (pbk)
ISBN13: 978-1-315-85875-3 (ebk)

Typeset in Sabon
by Keystroke, Station Road, Codsall, Wolverhampton
Printed and bound in Great Britain by Ashford Colour Press Ltd

Commo
Health nc

An essential and short guide for employees who need to know more about health and safety in the workplace without wanting to spend hours reading dozens of different documents. Whether it's for use alongside a training course or simply to brush up on your knowledge, it's perfect for equipping you with the principles of health and safety.

Friendly and accessible, this *Common Sense Guide* covers all the main aspects of health and safety in manageable chapters to provide you with the knowledge and understanding you need to look after yourself and others in the workplace.

- Suitable for the non-health and safety professional
- Includes questions at the end of each module to consolidate your health and safety knowledge
- Certificate offered to those who complete the exam at the end of the book and return to be marked externally.

Subash Ludhra is a past president of The Chartered Institution of Occupational Safety and Health (IOSH) and considered to be an expert in the field of Risk Management. Having qualified as an Occupational Hygienist, Subash Ludhra now manages Anntara Management Ltd; an international risk management and loss control consultancy business that operates in the UK and overseas.

[illegible] D HEALTH AND SAFETY

Co[illegible] ealth and Safety at Work
978-0-415-83[illegible] February 2014

Common Sense Guide to Fire Safety and Management
978-0-415-83542-8 February 2014

Common Sense Guide to Environmental Management
978-0-415-83541-1 February 2014

Common Sense Guide to International Health and Safety
978-0-415-83540-4 September 2014

Common Sense Guide to Health and Safety in Construction
978-0-415-83545-9 October 2014

Common Sense Guide to Health and Safety for the Medical Professional
978-0-415-83546-6 November 2014

Contents

Foreword

As the Chief Executive Officer of a large farming and fresh produce group, I am only too aware of the vital importance of getting health & safety right.

Management of health & safety, when implemented effectively, and driven forward with a passion to continuously improve, is a mission-critical element to running my companies in the correct legal, financial and, most importantly of all, ethical way possible.

It is a proven fact in our industry that a safe and well planned and organised company, is also a much more efficient one, which is viable and sustainable for the long term future.

When it has gone wrong for us, and it has, the negative impact on the business is almost too difficult to calculate, but even more important than that is the effect that it has on our people, in the short, medium and long term.

Therefore I need no convincing of the compelling case for good health & safety and a significantly important part of that is the need for high quality information, instruction, training and continuous learning and development of our people.

This new book is an excellent guide for non-health and safety professionals who need to know more about health and safety in the workplace. This guide supports basic training without being overly theoretical or specific, and offers an easy reference for learning the basics of health & safety.

It therefore provides a valuable new addition to our options to improve and develop our collective understanding of health & safety, and it follows that any improved learning gives us all a greater chance to take the actions needed today and tomorrow to avoid the devastation that personal injury causes.

It is an honour and a privilege to be asked by the author to write this foreword, and I wish you well as you strive each and every day to protect yourself and others.

It is a worthy objective.

John Shropshire – CEO
The Shropshire Group

Welcome

It is often said that we all know what is right and what is wrong, what we should and should not do. Surely it's just "common sense". Unfortunately common sense is not as common as we would like to believe. The definition of common sense is "the ability to behave in a sensible way and make practical decisions". Most individuals and employers will at some time do things that are not sensible or practical and some will do this regularly.

This guide has been developed to help improve your knowledge of health and safety in the workplace in a light-hearted way. It is designed to further heighten the common sense element of your existing knowledge.

Although the guide refers to UK legislation/best practice the principles are applicable internationally. However, you should answer all questions within the guide, based on the UK legislation.

By taking the time to improve your knowledge and learn more about health and safety in the workplace you can:

- **avoid having an accident**
- **help prevent accidents occurring to others**
- **make your workplace safer**
- **make your home safer.**

Every year, hundreds of people are killed while at work in accidents, and thousands more are killed or suffer from longer-term work related ill health. The cost to industry runs into billions of pounds and as a result everyone suffers.

Most importantly if you are injured as a result of an accident or suffer from work-related ill health you and your family are likely to suffer.

DID YOU KNOW?

- On average 170 people are killed each year in accidents while at work.
- On average over 12,000 people are killed each year as a result of work-related ill health (including asbestos-related fatalities).
- An estimated 2 million people in Great Britain each year suffer from an illness which they believe was caused or made worse by their current or past work.
- An estimated 27 million working days are lost overall each year (23 million due to work related ill health and 4 million due to workplace injury).
- By working through this accessible guide you will learn more about health and safety in the workplace and as a result you will be better placed to recognise the hazards and dangers in any workplace or even your own home 24 hours a day.

Injured man

How this guide works

AIMS

The aim of this guide is to provide you with a basic understanding of:

- the need to manage health and safety
- health and safety law and enforcement
- health and safety legislation
- accidents/incidents and what causes them
- hazards in the workplace and
- proactive measures that can be taken to help reduce accidents to ensure that you can work in a manner that is safe for you and your colleagues at all times.

OBJECTIVES

By the time you finish this guide, you will be able to:

- identify hazards within your place of work
- define hazard and risk
- help your employer improve health and safety standards at your place of work
- assess risks
- understand Civil and Criminal law relating to health and safety
- lift and move loads in a safer way
- recognise and participate in a number of your employer's proactive safety measures
- recognise safety signs and understand their meaning
- avoid taking unnecessary risk
- work safely with electricity
- know when your environment is too noisy
- know what a reportable accident is and who to report it to

- have a greater understanding of health and safety and its importance in your everyday work and home life.

How to complete the guide

Before you start to complete this guide please read the notes below in order to ensure that you get the most out of your training.

WHAT DO YOU NEED TO COMPLETE THE GUIDE?

You will need:

- a quiet, cosy environment that allows you to relax and make notes;
- a pen and paper to make notes;
- a desire to learn and improve your knowledge of health and safety.

GUIDANCE ON LEARNING

This guide has been produced to help you learn more about health and safety in your workplace and to reduce the likelihood of you or your colleagues having an accident or suffering from ill health. The guide allows you to complete your studies at your own pace with the support of your manager or supervisor. However we recommend that you complete the guide within four weeks.

Throughout the guide there are simple questions designed to help you test your subject knowledge and learn.

If you cannot answer the questions please read the relevant topic again to refresh your memory. If you are still in doubt please speak to your line manager who will be able to assist you.

When you have completed the guide and the exercises, you may wish to complete the examination (30-question multiple-choice exam paper on pages 101–110). On achieving the required pass mark (75%) you will receive a certificate to confirm that you have completed the guide and passed the associated examination.

HOW TO USE THE GUIDE

The guide is divided into modules. We recommend that you complete one module at a time in full, starting with module one, progressing sequentially through to the last module. You do not have to complete the guide in one sitting. A lot of information is provided and you may learn more effectively by tackling the modules in bite-sized chunks.

The modules are designed to take you through a specific learning pattern to help you learn. Each module contains questions to make you think about your own job and workplace. There are also questions at the end of each module to test your knowledge and understanding of it. The answers can be found on pages 91–6 of the guide. There may be times when you feel you need help and support in completing the guide. Should this be the case please speak to your line manager.

WHEN DO I GET MY CERTIFICATE?

Once you have completed the guide and the examination paper, the paper will be marked and on achieving the required pass mark a certificate pdf will be emailed to you as soon as possible.

Remember the certificate only confirms that you have completed the guide and passed the associated exam. The real benefit to you will come from your improved knowledge and ability to identify hazards and reduce the risk of having an accident at work or in the home. **Good Luck!**

Winning trophy

MODULE ONE

Health and safety law and enforcement

£1,000,000 FINE FOR HIGH STREET RETAILER!

. . . FOR FAILING TO ENSURE THE HEALTH, SAFETY AND WELFARE OF THEIR EMPLOYEES AND OTHER PERSONS NOT IN THEIR EMPLOYMENT

Welcome to module one. In this module you will learn about the importance of health and safety and the Health and Safety at Work Act.

WHAT IS HEALTH AND SAFETY?

Health and safety is about the measures necessary to control and reduce risks to an acceptable level, to ensure the health and/or safety of anyone who may be affected by the activities of people at work.

> "IN REALITY A WORKPLACE CAN NEVER BE 100% RISK FREE."

However, adequate controls and systems need to be in place that are acceptable to the employer, employees and the enforcement authorities.

Any instruction given to you concerning health and safety is for your own well being and must be adhered to at all times.

Remember successful health and safety management is a two-way commitment between employers and employees and the systems can only be successful when both parties co-operate.

Health and safety management revolves primarily around the term REASONABLY PRACTICABLE (i.e. as an employer, did the company do what was reasonably practical to safeguard their employees or others affected by them?).

Employers and employees must work together

This is a balancing act of RISK (the probability of an event occurring and the likely consequences if it does occur) against COST (this may be in terms of benefit, money, time, effort, latest technology etc.).

If the RISK relates to slipping in the kitchen and the COST of removing/replacing the slippery surface is very high, then it may be acceptable to minimise the risks in other ways – through warning signs, coatings, the provision of appropriate safety footwear and so on. Generally the greater the risk the less important the cost element becomes.

Reasonably practicable

WHY ADDRESS HEALTH AND SAFETY?

1. To comply with statutory legislation

You and your employer are required to comply with the requirements of the Acts and Regulations relevant to your employer's business (this will be covered in more detail later). Examples of Acts and Regulations include:

- The Health and Safety at Work Act
- The Control of Substances Hazardous to Health Regulations
- The Food Safety Act
- The Food Safety (General Food Hygiene) Regulations
- The Management of Health and Safety At Work Regulations
- The Personal Protective Equipment At Work Regulations
- The Manual Handling Operations Regulations
- The Health and Safety (Display Screen Equipment) Regulations
- The Workplace (Health, Safety and Welfare) Regulations
- The Provision and Use of Work Equipment Regulations
- The Health and Safety (Consultation with Employees) Regulations
- The Reporting of Injuries, Diseases and Dangerous Occurrences Regulations
- The Environmental Protection Act
- The Health and Safety (First Aid) Regulations.

This is only a small selection of the Acts and Regulations that companies have to comply with.

2. To comply with internal company policies

Responsible employers have internal health and safety policies and procedures. These policies and procedures set out their own internal standards, which in some cases may exceed the requirements of the relevant regulations. You, as an employee, agree to comply with these as part of your terms and conditions of employment.

3. Moral obligations

To reduce the number of accidents and incidents and to safeguard you the employee. No employee arrives at work expecting to have an accident. Therefore by increasing your own awareness of health and safety you can help reduce the likelihood of accidents and incidents occurring and safeguard your well being.

4. It's cost effective

All accidents cost money, for example by staff working overtime, the cost of employing agency staff, loss of production, failure to meet service requirements and extra pressure on the staff left to cover absenteeism. However, the real cost is the damage to the injured employee's well being and their suffering after the accident. This also affects their family and relations who are sometimes overlooked after an accident occurs.

5. There is a business need to do so

To maintain your employer's good reputation, more and more companies require their suppliers to have good health and safety systems in place and to be able to demonstrate the promotion of a good, safe working environment.

"A POOR SAFETY REPUTATION COULD COST YOUR EMPLOYER'S BUSINESS DEARLY."

Failing to manage your health and safety can cost you your business

THE HEALTH AND SAFETY AT WORK ACT 1974

The Health and Safety at Work Act was introduced in 1974 to help protect the large number of employees who, at the time, were not covered by any health and safety legislation. Its aims were:

- **a)** to secure the health, safety and welfare of persons at work;
- **b)** to protect other people from health and safety risks caused by their work activities, for example contractors/visitors on site;
- **c)** to control the storage and use of explosive, flammable and dangerous substances (this has developed to become the Control of Substances Hazardous to Health Regulations);

d) to control atmospheric emissions of certain substances that could prove to be harmful (which has now developed into the Environmental Protection Act).

More importantly it set out clear definitions of responsibilities for employers, employees, the self-employed and suppliers/manufacturers.

EMPLOYERS' DUTIES

Employers have a duty under the Health and Safety at Work Act to:

- provide and maintain plant and systems of work that are safe and without risk to health;
- ensure safety and absence of risks to health in connection with the use, handling, storage and transport of articles and substances;
- provide information, instruction, training and supervision to ensure the health and safety at work of employees;
- provide you with a place of work which is in a condition that is safe with respect to access and egress;
- provide and maintain a working environment with adequate welfare facilities such as toilets/washing facilities;
- ensure that persons not in their employment, that is, visitors, contractors and so on, are not exposed to risks to their health and safety.

Obey the law and your duties

EMPLOYEES' DUTIES

As well as employers you as an employee have a legal and moral obligation:

- to care for the health and safety of yourself and of other persons who may be affected by what you do or do not do;
- to co-operate with your employer by complying with internal and external policies imposed for work activities to ensure your employer is able to fulfil his/her legal obligations;
- not to intentionally or recklessly interfere with or misuse anything provided in the interests of health, safety or welfare. Your employer may have provided safety equipment and procedures to help reduce risk in the workplace and your co-operation is required at all times.

A failure to meet your legal obligations could lead to internal disciplinary action and/or external enforcement action directly against any employee who is found to have broken a health and safety rule that results in injury or property damage.

Tampering with or damaging safety equipment, for example machinery guarding, personal protective equipment, fire extinguishers and so on, could lead to dismissal by your employer or prosecution by external enforcement authorities.

"REMEMBER: WORKING SAFELY MUST BE A CONDITION OF EMPLOYMENT."

DUTIES OF THE SELF-EMPLOYED

Under the Health and Safety at Work Act the self-employed have a duty to ensure that their work activities do not endanger themselves or others.

SUPPLIERS/MANUFACTURERS

Suppliers and manufacturers of articles and substances have a duty to:

- ensure that the product designed or constructed is safe when properly used;
- test or have the product tested to ensure that it is safe;
- provide information and instructions for the user.

HEALTH AND SAFETY LAW

There are two important areas of law in health and safety:

Civil Law Tends to deal with the area of compensation awarded or claimed when incidents occur or something goes wrong.

County / magistrates' court

Criminal Law Deals with the punishment awarded by the courts when an employer/employee breaks the law.

LAW ENFORCEMENT

The policing and monitoring of health and safety in the UK is carried out and enforced by Factory Inspectors and Environmental Health Officers (EHOs).

They have a wide range of powers, which include:

- visiting a place of work at any reasonable time
- carrying out investigations and examinations
- dismantling and removing equipment
- taking samples of products
- taking photographs and viewing documents
- taking statements from key personnel
- requiring assistance.

Their primary function is to help to ensure employers are complying with the relevant regulations and safeguarding their employees and others affected by their business. However, whilst in their enforcement role they can also:

- Issue an **Improvement Notice.** This means something is not safe and they want it made safe; they will stipulate a reasonable time period to allow the employer to comply.
- Issue a **Prohibition Notice.** This means that something is so dangerous that the employer/employee must stop doing it immediately. The employer must then make it safe and be able to demonstrate to the enforcement body that it is safe before recommencing the activity.
- **Prosecute** employers and employees. This is generally only carried out when there is very strong evidence to

suggest that an individual/company deliberately did something they should not have been doing.

PENALTIES

In a magistrates' court the maximum fine for a health and safety offence is **£20,000** (per offence) and/or a **six-month prison sentence.**

In a crown court the maximum fine is **unlimited** and in addition a **two-year** prison sentence can be imposed.

Note: The Health and Safety Executive's inspectors can also charge for their time under the fee for intervention (FFI) scheme where they identify material health and safety breaches.

HEALTH AND SAFETY ROLES AND RESPONSIBILITIES

As well as the duty imposed on employers and all employees under the Health and Safety at Work Act, within your organisation a number of individuals may have been given specific health and safety duties and responsibilities by virtue of their job functions. These individuals help to ensure that safety is taken seriously and is properly managed.

Q

Do you know who has specific health and safety responsibilities in your workplace? (Write your answer here: if the answer is no then find out.)

____

____

EXERCISE 1

Self-assessment questions

1. What is the maximum fine you can receive at a magistrates' court?

a) £10,000
b) £15,000
c) £20,000
d) £30,000

2. What are the two important types of law?

a) Civil & Matrimonial
b) Criminal & Peaceful
c) Civil & Criminal
d) Peaceful & Matrimonial

3. Can an enforcement officer take equipment away from your workplace?

a) Yes
b) No

4. List four reasons for employers to address health and safety:

a) ....

b) ....

c) ....

d) ....

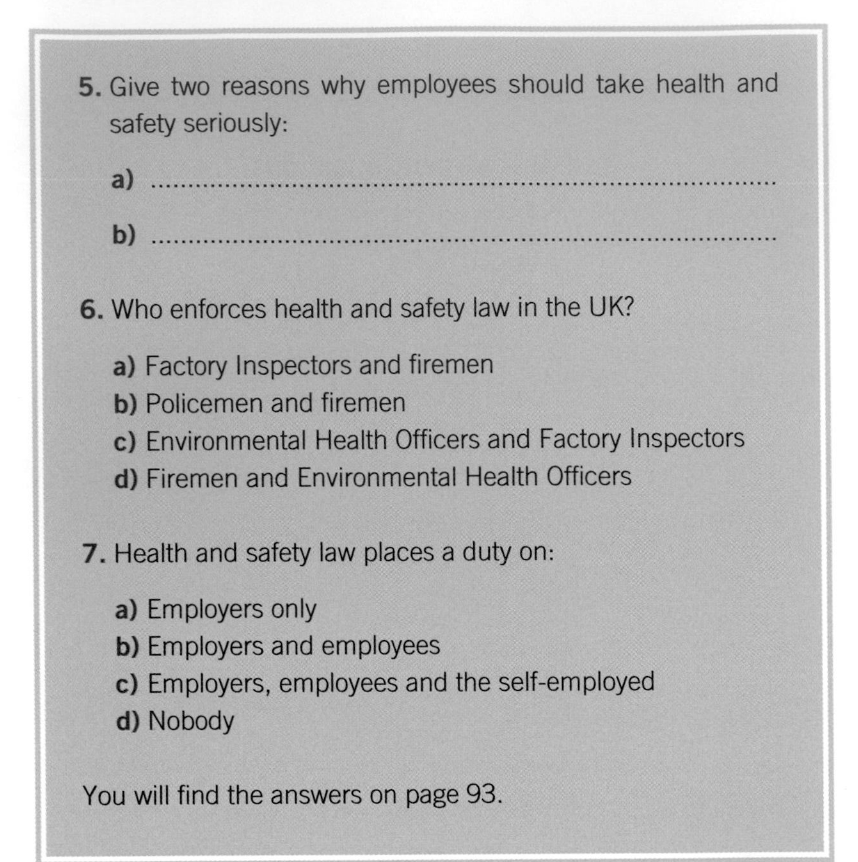

5. Give two reasons why employees should take health and safety seriously:

a) ....

b) ....

6. Who enforces health and safety law in the UK?

a) Factory Inspectors and firemen
b) Policemen and firemen
c) Environmental Health Officers and Factory Inspectors
d) Firemen and Environmental Health Officers

7. Health and safety law places a duty on:

a) Employers only
b) Employers and employees
c) Employers, employees and the self-employed
d) Nobody

You will find the answers on page 93.

MODULE TWO

Health and safety regulations

COMPANY DIRECTOR FINED AFTER FAILING TO CONTROL ASBESTOS

Welcome to module two. In this module you will learn more about specific health and safety regulations that are likely to apply to your place of work.

SAFETY SIGNS USED AROUND YOUR WORKPLACE

You will no doubt have seen signs on display in your own place of work and/or other places of work. Signs cost money and give an important message. They are not displayed to make the workplace look pretty.

Signs are produced in four key colours to ease identification. Each colour gives a particular message, which should always be adhered to.

Always know where the signs are, what they mean and most importantly ensure that you comply with their requirements

COLOUR	MEANING	EXAMPLE
BLUE	Mandatory (must do)	Wash your hands
RED	Prohibition (must not do)	No Smoking
YELLOW	Warning of hazards	Caution wet floor
GREEN	A safe condition	Fire escape route/First aid box

Q

Think about some of the signs in your place of work. Can you describe three?

NOISE

What is noise?

- Noise is generally unwanted sound.
- Noise levels are measured in decibels (dB).
- The human ear's performance generally reduces with age. This process can be hastened by exposure to noise. Exposure to very loud noises can cause instant damage, even deafness. However, damage is typically caused by exposure to noise over a longer period of time.
- The Noise at Work Regulations were introduced to protect people from excesses of noise.

Two key action levels were introduced. These are:

80dB(A)

If noise levels exceed this (averaged over 8 hours) then your employer is required to:

- carry out a risk assessment (by a competent person);
- make reasonable attempts to reduce noise levels using control measures to below this point. If he/she cannot then he/she is required to inform you, his/her employees, and offer hearing protection to you if you wish to use it.

85dB(A)

If noise levels exceed this then hearing loss is likely. If the levels cannot be reduced through engineering controls then the employer has to:

- carry out a risk assessment (by a competent person);
- provide suitable hearing protection;
- ensure that you wear the protection provided;
- mark out hearing protection zones.

The A-weighting is an expression of the relative loudness of sounds in air as perceived by the human ear, as the ear is less sensitive to low audio frequencies compared to high audio frequencies.

The control measures available to your employer to help prevent hearing damage include:

- the use of suitable equipment
- the selection of new, quieter equipment
- planned maintenance of equipment

- enclosing the noisy equipment with sound insulating material
- enclosing the workforce in sound-insulated booths
- reducing the time employees are exposed to the noise
- providing suitable personal protective equipment (PPE)
- marking out noisy areas as Ear Protection Zones.

TYPICAL NOISE LEVELS (dB(A))	
Jet aircraft (taking off at 25m)	140
Chainsaw	120
Heavy traffic	80
A busy office	60
A typical library	20

Ear protection must be worn

This is a mandatory sign and must be obeyed

As a guide, if you are standing two metres away from a colleague and cannot hear them when trying to have a normal conversation the environment is probably too noisy.

Remember once hearing is damaged it may never be repaired.

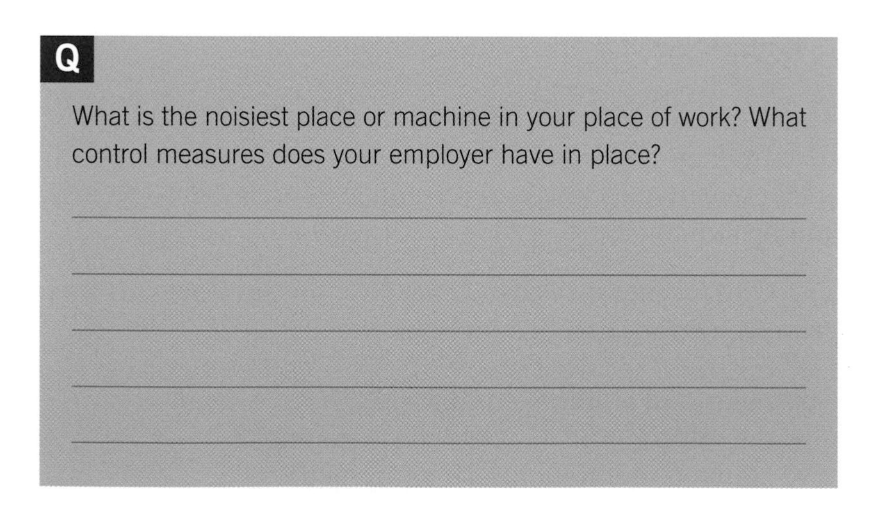
Q

What is the noisiest place or machine in your place of work? What control measures does your employer have in place?

ELECTRICITY

Overloaded plug socket

Electricity is used in almost all workplaces and is often referred to as the invisible killer as it can kill without being seen.

Electricity can cause:

- electric shock
- electrical burns
- death
- fires.

The Electricity at Work Regulations were introduced to protect people from dangers associated with electricity and to protect equipment from excesses of electricity.

Your employer must ensure that the electrical equipment on site is:

- safe to use
- properly maintained/tested
- easily isolated from the mains supply
- only worked on by competent people
- protected from excesses of current
- protected from adverse weather conditions or other hazards.

If using electrical equipment you must:

- ensure that the equipment you use is properly maintained
- be competent to use the equipment
- visually inspect it for signs of damage before use.

Remember it is the electrical current (measured in amps) that kills, not the voltage. A very small current passing through your body could kill you.

When using portable appliances plugged into the mains always ensure that they are protected by using an in-line circuit-breaking device and never handle electrical tools with wet hands.

In any case of a suspected electric shock:

- call for help;
- switch off the power supply if possible and safe to do so, then move the person away from the electrical source using a non-metallic object (do not touch them directly if there is a risk of electrocution to you);
- if the casualty is breathing place them in the recovery position and call for medical attention;
- if the casualty is not breathing call for help and attempt resuscitation;
- never touch the casualty with bare hands unless the power supply has been isolated;
- only give first aid if you have been trained to do so.

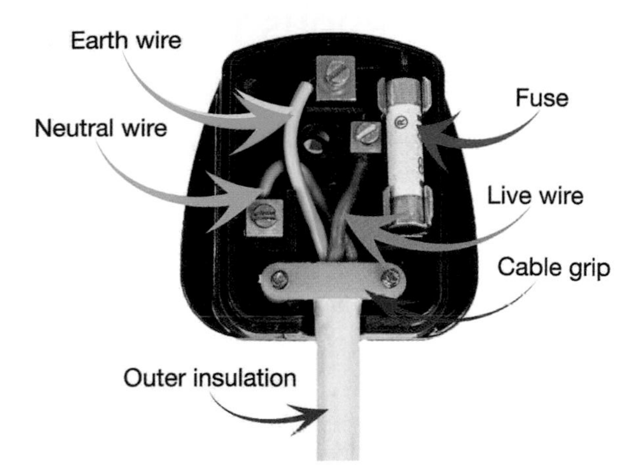

Ensure plugs are wired correctly

Q

Do you use any electrical equipment at work or in the home? What precautions do you take prior to and during use?

SUBSTANCES HAZARDOUS TO HEALTH

What are substances hazardous to health?

Any materials, mixtures or compounds used at work or arising from work activities, which have the potential to cause harm to people's health in the form in which they occur in the work activity.

Poor labelling on containers

Examples include:

- acids used in factories for specialist cleaning
- toners used in photocopiers within offices
- correction fluid used within offices
- bleaches used as cleaning agents
- tints used in hairdressing salons
- glues and solvents used in floor laying.

They can be:

- liquids, solids, dusts, powders, gases or vapours.

They may typically be:

- toxic, corrosive or irritant.

They can cause damage by:

- coming into contact with the skin and eyes;
- entering the body through cuts in the skin;
- being breathed in;
- entering the body through the mouth.

Cleaning chemical

The Control of Substances Hazardous to Health Regulations (COSHH) require your employer to protect you from exposure to hazardous substances by:

- carrying out an adequate assessment of the hazards;
- preventing/controlling your exposure to hazardous substances.

When carrying out the assessment your employer must:

- identify hazardous substances in the workplace;
- identify who is at risk;
- obtain the relevant material safety data sheets (MSDSs);
- evaluate the risk.

Once the assessments are complete he/she must:

- eliminate the risk of exposure or introduce appropriate and effective control measures;
- maintain the measures put in place;
- monitor the effectiveness of the control measures;
- record the assessments;
- ensure he/she informs you of all risks and provide suitable instruction, training and supervision.

Control measures that can be applied to reduce risk include:

- elimination
- substitution
- prevention of exposure
- limiting exposure time
- providing local or general ventilation
- improving housekeeping
- training
- providing health screening
- providing appropriate personal protective equipment (PPE).

When handling substances hazardous to health always:

- ensure you have been made aware of the hazards and risks;
- store them in correctly labelled containers;
- wear any protective equipment provided by your employer;
- treat them with respect.

Q

Which substances hazardous to health do you use at work or in the home? What precautions do you take?

FIRST AID

First aid signage

No matter how safe your place of work is there is always the potential for someone to have an accident or fall ill. This is why first aid forms a vital part of any employer's health and safety management system.

The Health and Safety (First Aid) Regulations were intended to ensure that in the event of an incident a qualified person would be able to preserve life and prevent the deterioration of someone who has fallen ill or is the victim of an accident.

Your employer must assess the risks within your workplace and provide adequate cover.

In principle, your employer is required to have at least one fully qualified first aider per 50 employees (or emergency first aider as appropriate). That said adequate cover must be provided during normal working hours taking into account shift patterns, holiday and sickness cover, the size of the location and hazards present. High-risk environments may require further specialist training for the first aider. In very low-risk environments an appointed person may be sufficient. It would be their responsibility to provide basic medical cover, call for medical assistance and maintain any medical supplies on site.

It is important that you ensure you know who your first aiders are, where their normal place of work is and how to contact them. In the event of an emergency, it could help save lives. If you are not sure ask your manager.

First-aid kits must contain as a minimum: individually wrapped plasters, sterile eye pads, triangular bandages, sterile dressings and safety pins. First aiders are not authorised to issue or apply creams, pills, lotions, sprays or any other medicines.

Q

Do you know who your local first aider is, where they are based and how to contact them?

MANUAL HANDLING

Manual handling is the lifting, lowering, pulling, pushing or moving by hand or bodily force of an object or load. It is a function you perform each and every day from brushing your teeth to picking up a shopping bag.

Some job functions require more manual handling than others and it's not just the weight of a load that can cause problems. In fact the incorrect manual handling of any load can cause injury.

In the UK millions of working days are lost each year due to pains, strain and other injuries to the back. Manual handling injuries make up about a quarter of all injuries reported to the enforcing authorities.

Injuries can occur from a single event or may develop over a period of time.

The types of injury suffered from manual handling are not confined to the back and can include cuts, hernias, wrenched shoulders, crushed feet or fingers, fractures and bruises.

The Manual Handling Operations Regulations were introduced to help reduce the number of accidents occurring within the workplace. In simple terms the regulations say that if manual handling can be avoided then it should be. If it cannot be avoided then your employer is required to carry out a risk assessment and introduce control measures to minimise the risk of injury to you.

The assessment must take into account:

- the **Task**, that is, what you are actually required to do;
- **Individual capability**, that is, what you are capable of handling;
- the **Load**, that is, what it is you are required to handle;
- the **Environment**, that is, your surroundings.

Example of mechanical aids

Always remember the following procedure:

1. Examine the load – can the manual handling activity be avoided by redesigning or changing the task or with the use of mechanical aids?
2. Are you physically capable of handling the load or is assistance required?
3. Is protective clothing required to protect your hands, body or feet?

When lifting a load:

1. Always face the way you are walking or working – avoid twisting your trunk and overstretching at all times.
2. Always position your feet so that they are approximately the width of your hips apart, with one foot slightly in front of the other and flat on the floor to provide a stable, balanced stance. Once the load has been lifted, the weight may be transferred to the front foot.

3. Always ensure you have a firm grip on the load even if it means using gloves to prevent the load slipping.
4. Your back should be kept straight to maintain it in its natural and strongest position. To get down to the load, the knees and the ankles should be bent and the load raised gradually using the thigh and leg muscles.
5. The head should be kept up and the chin well in, as this helps to keep the spine in its natural upright position.
6. Arms should be kept as close to the body as possible, which helps to retain balance.
7. The body should be used to counterbalance the weight of the load.
8. If it is a team lift then one person should control the lift so that it is even and together.

When carrying a load always ensure:

1. You can see where you are going. Loads which extend to head height and which obstruct your vision are dangerous to you and to other persons nearby – if this is the case you should either use a mechanical aid (sack barrow or pallet truck) or seek assistance from a colleague.
2. Your route is clear of any obstructions or slippery areas.
3. Particular care is taken when going around corners or negotiating stairs.
4. Sudden movements and twisting of the spine is avoided – face the way you are walking.
5. That if the load cannot be set down in the required position it is put down temporarily and re-lifted when the position is clear.
6. You do not try to change your grip while carrying – if your grip is slipping put the load down and start again.

When putting down a load always ensure:

1. Your back is kept straight and in its natural position. If the load is to be lowered, the thigh and leg muscles must be used.
2. Your head is up and your chin tucked in.
3. You maintain a proper palm grip, but beware of trapping the fingers and hands beneath the load.
4. You use your body weight to counterbalance the load.
5. You keep your arms close to your body. Setting down and stacking should be only as high as it is possible to reach with the elbows tucked into the sides. Do not over-reach.
6. You maintain a firm, balanced stance, with your feet as close as possible to the centre of gravity of the load, and flat on the floor. If the load you are moving is too heavy or bulky always seek assistance, which can be a colleague of similar height and build (one person should call the signals) or mechanical aids.

PERSONAL PROTECTIVE EQUIPMENT (PPE)

Personal protective equipment

Personal protective equipment is any device or piece of equipment held or worn that can provide protection against one or more hazards.

When used correctly PPE can be used as an effective control measure against hazards. However:

> "WEARING PERSONAL PROTECTIVE EQUIPMENT DOES NOT REDUCE THE HAZARD AT ALL. THE HAZARD REMAINS THE SAME EVEN THOUGH THE RISK TO YOU IS REDUCED."

As such it must always be used as a last resort when all other control measures have been exhausted. The Personal Protective Equipment Regulations require your employer to carry out a risk assessment and identify when and where the wearing of PPE will be necessary.

When using PPE (which he/she must provide at no cost to you) your employer must take into account the following:

- its fit, that is, is it the right size/shape for you?;
- its suitability against the risk;
- the ergonomics of the PPE, that is, how well it is designed for the individual user;
- any increase in risk to you or others as a result of using the PPE;
- its ongoing cleaning/maintenance;
- replacement procedures;
- storage facilities;
- adequate training to ensure you understand why it needs to be worn, the benefits it will provide to you, how it is to be maintained and stored and how you obtain replacements.

When provided, you as the employee are required to wear it and take reasonable care of it.

Questions to ask yourself when in your workplace:

- Is there a need for PPE for the job I am doing?
- Do I need protecting against any hazards?
- If I use PPE is it suitable and does it serve its purpose?
- Is the PPE in good condition and cleaned regularly?
- Does it fit correctly and is it comfortable to work in?
- Is it readily available?
- Can I get replacements?
- Have I been trained on how to use the PPE? (on more complex equipment recorded evidence of training must be documented).
- Does the PPE cause me any problems while I am working (heat rash or allergies)?
- Where do I store it when I have finished with it?

EXAMPLES OF PPE

Body part protected	Type of PPE available
Body	Overalls, coveralls, tabards
Feet	Shoes, boots
Hands	Gloves, gauntlets, wrist cuffs
Ears	Ear plugs, muffs
Eyes	Goggles, glasses
Face	Visor, hood, shield
Head	Bump caps, helmets
Skin	Barrier creams
Lungs	Breathing apparatus, respirators
Legs	Knee pads, leggings

Q

Do you or any of your colleagues wear PPE? If so what is it worn for and how do you get it replaced?

____

____

____

____

____

____

YOUR WORKPLACE

Typical office layout

The Workplace (Health, Safety and Welfare) Regulations were introduced to update, strengthen and reinforce many of the elements within the previous Factories Act of 1961.

Your employer is required to provide you with adequate facilities and ensure the workplace is safe and healthy.

Ventilation

Air must be provided either by natural (e.g. an open window) or mechanical means (e.g. a ducted-fan-assisted system).

Temperature

The workplace temperature must be at least 16°C (or at least 13°C where the work involves physical effort). If you work outside or in areas where it is unreasonable to maintain the minimum temperature or in areas that require temperatures to be kept high or low then your employer may have to provide you with specialist clothing or make other suitable arrangements.

Lighting

Your employer is required to ensure that there is suitable and adequate lighting for the tasks that you perform.

Stairways

Should be kept clear and well lit. Guardrails must be fitted to open stairways.

Workstations

Your employer must ensure that you have adequate space to perform your job safely.

Furniture

Must be suitable for the tasks being performed.

Floors/walkways

Must be kept clear and in good repair to prevent slips, trips or falls.

Housekeeping

The workplace must be kept clean and tidy. Equipment should be correctly stored and not left lying around.

Example of a hazard sign

Toilets

Adequate numbers of urinals and water closets must be provided.

Washbasins

Based on the number of male/female employees present. They must be suitably maintained and cold and warm water must be available.

Rest Areas

May be required depending on the nature of your business.

WORK EQUIPMENT

Work equipment is defined as any machine or hand tool used at work. From this definition it is clear that almost all workplaces will have work equipment present. Although work equipment is generally there to make tasks easier, quicker or more efficient, if used incorrectly or without training, the work equipment can be dangerous and cause injury.

Under the Provision and Use of Work Equipment Regulations (PUWER) your employer is required to:

- provide suitable and safe work equipment;
- provide suitable and sufficient training on the correct use (including risks and precautions necessary) of the equipment for all authorised operators of the equipment.

There are a number of specific dangers associated with work equipment. These include:

Entanglement

Loose clothing or jewellery worn by you could become entangled within machinery.

Entrapment

Some machines can entrap parts of your body within their moving parts. Fingers, limbs and even the whole body can be pulled into the machine and crushed.

Emergency stop button

Gas cylinders and welding equipment

Contact

Moving parts of a machine could cause severe injuries to limbs or burns to the skin.

Ejection

Moving or rotating machines can throw out objects unexpectedly, causing you injury.

Impact

You could be struck by a moving machine or an object being worked on by a machine.

Greater safety when using work equipment can be achieved by:

- buying and only using equipment that has been designed and constructed to remove foreseeable dangers;
- ensuring that equipment is used in a safe place not giving rise to danger;
- guarding dangerous parts of equipment to prevent injuries. Guards are designed and fitted to equipment to protect you and the equipment. Never tamper with guards, try to override them or operate a machine with the guards removed;
- never wearing loose clothing or jewellery around machinery;
- covering or tying long hair to prevent entanglement;
- ensuring adequate lighting so that everything can be clearly seen;
- ensuring that you wear any personal protective clothing deemed to be necessary;
- switching off and isolating the equipment when not in use and preventing unauthorised use;
- ensuring that the area around the work equipment is clean and tidy to prevent things or people falling;
- reporting all faults or suspected faults to your manager so that they can be investigated and where necessary rectified;

- only using work equipment if you are trained, competent and authorised to do so;
- never distracting other people who are using work equipment.

Abrasive wheel with guard missing

When using hand-tools the following rules apply:

- Visually inspect the tools for signs of damage prior to use and report any defects.
- Always use the correct tool for the job. Do not attempt to make the tool fit the job.
- Ensure that the tools are in good condition.
- Use the tools in the correct way.
- Never use tools you have not been trained to use.
- Take out of service any broken or damaged tools identified.
- Work equipment can only ever be as safe as the operator using it.

DISPLAY SCREEN EQUIPMENT

Computer workstation

The Display Screen Equipment Regulations were introduced to protect users of display screen equipment from the effects of:

- **Musculo-skeletal pains** – these are the aches and pains that can occur in the wrists, arms, neck and upper/ lower back as a result of poor posture or workstation design.
- **Eye strain** – this can occur from poor screen adjustment, bad lighting or even from the use of incorrect glasses, and is usually associated with headaches.

- **Mental fatigue** – this can occur as a result of poor workload planning, or limited understanding of the software or hardware being used. Your employer is required to assess all workstations (which includes the desk, chair, keyboard and working environment) to ensure that they comply with the requirements of the regulations and ensure the well-being of users of the equipment. If you are a user of display screen equipment you may be asked to assist in the completion of the assessment form as part of the process.

Your workstation should meet the following criteria:

The screen

- Is the contrast and brightness adjustable?
- Are the characters of adequate size, stable and not flickering?
- Can the screen be easily swivelled and tilted to suit you?
- Is the screen free from glare or reflections?

The keyboard

- Is the keyboard detached from the bulk of the processor?
- Is the keyboard tiltable?
- Are all the keypad symbols legible?
- Is the surface of a matt finish?

Space

- Is there sufficient space to work comfortably?

Lighting

- Is there glare or a reflection on the screen from lighting or windows?
- Can the screen be moved to avoid the glare?
- Are there blinds to prevent reflections from sunlight?

Noise

- Is the workstation positioned alongside noisy machinery (printers etc.)? If so can they be moved or covered?

Environment

- Is the workplace atmosphere so hot or cold as to make the user uncomfortable?

Cables

- Are all cables in a sound condition, not frayed or causing a trip hazard?

Seating

- Are there five castors on the chair?
- Can the back be adjusted to support the lower part of the back?
- Can both feet be placed flat on the floor? If not, is a footrest provided?
- Can the seat height be adjusted?

Work surface

Is it of low reflection and large enough to accommodate any other piece of equipment needed, i.e. document holder, calculator, telephone?

If your workstation does not comply with the above or should you have any concerns, please contact your manager for guidance.

USEFUL GUIDELINES ON HOW TO USE YOUR WORKSTATION

Seating

- Avoid slouching, sit upright and maintain the natural curve of your back.
- Adjust the backrest of your chair to support your lower back.
- Sit back in your chair to gain full support.
- Arrange your workstation to help maintain an upright posture.
- Don't sit at your work place for long periods of time – try to plan your workload to have breaks away from the screen by doing other tasks.
- Use a footrest if your feet cannot sit comfortably on the floor.

Upper body

- Adjust your seating so that your forearms are in a horizontal position.
- Align your hands with your forearms so you are working with straight supported wrists.
- Adjust the angle of your screen to suit your own sitting position and eye level.

Vision

- Position your screen so that reflections from windows (sunlight) or internal lighting are not causing problems on your screen. Window blinds may also be necessary.
- Position your screen so that the need to move your head and neck is minimised.
- If you use a document holder position it alongside your screen to reduce movement of the eyes, head and neck.
- Adjust the screen to minimise the reflections from the lights, windows or other bright surfaces.
- Regularly clean your screen with the equipment provided.
- If you are identified as a user of display screen equipment then you may be entitled to an eye test paid for by the company.

In order to achieve the above it may be necessary to re-arrange furniture.

Remember, the person using the equipment before you may have been of a different size and height.

Incorrect use of your workstation and equipment can lead to physical discomfort (backaches, wrist problems etc.), eye strain (blurred vision, headaches) and stress.

If in doubt, speak to your supervisor or manager for assistance in making adjustments to your workstation.

Your PC may have connections to a network system. Do not unplug any cables unless you have checked with your system administrator.

Laptops

- If you use a laptop for work than it should ideally be used with a docking station if being used for long periods of time.
- Do not use laptops on your lap.
- Try and use external mouse devices.
- Ensure it is used on a suitable desk so that you can maintain a reasonable body posture.
- Remember that carrying a laptop around constitutes manual handling so think about how it is being carried.

EXERCISE 2

Self-assessment questions

1. What are the four colours used for safety signs?

   a) ....

   b) ....

   c) ....

   d) ....

2. What are the two key action levels under the Noise at Work Regulations?

   a) 85 & 95dB(A)
   b) 80 & 85dB(A)
   c) 75 & 85dB(A)
   d) 80 & 90dB(A)

3. Can exposure to one very loud noise cause deafness?

a) Yes
b) No

4. List four different health and safety control measures that can be applied to reduce risk.

a) ....

b) ....

c) ....

d) ....

5. Which muscles should be used when lifting a load?

a) Arm muscles
b) Leg muscles
c) Back muscles
d) Stomach muscles

6. What is the purpose of first aid?

....

....

....

7. List two things that must be assessed when carrying out a VDU assessment.

a) ....

b) ....

You will find the answers on pages 93–4.

MODULE THREE

Accidents and incidents

THE INTERNATIONAL LABOUR ORGANISATION (ILO) ESTIMATES THAT SOME 2.3 MILLION WOMEN AND MEN AROUND THE WORLD SUCCUMB TO WORK-RELATED ACCIDENTS OR DISEASES EVERY YEAR.

Welcome to module three. In this module you learn more about accidents and incidents, how they occur and what needs to be done when they do occur.

ACCIDENTS AND INCIDENTS

What is an accident/incident?

These are unplanned/uncontrolled events which have led to, or could lead to, damage or injury to individuals, property, plant or any other loss to an organisation.

Accidents and incidents occur as a result of unsafe acts and unsafe conditions which are affected by:

ORGANISATIONAL FACTORS

These are factors relating to the organisation you work for, for example:

- the safety culture within the organisation;
- management commitment to safety.

HUMAN FACTORS

These are factors that may relate to you, for example, your attitude, the training you have had, any disabilities you may have, your health and/or fitness.

OCCUPATIONAL FACTORS

These are factors relating to the work you do and the environment you do it in, for example:

- If you deliver products there will be a greater risk of muscular strain due to the manual handling involved.
- If you work in a kitchen you are more at risk from knife injuries and burns from touching hot surfaces.
- High noise levels in a factory can damage your hearing.
- Poor ventilation within a laundry or kitchen can cause heat stress.
- Poor lighting within an office can cause headaches and eyestrain.
- Working in a dusty atmosphere such as a commercial bakery can cause breathing problems.

Once things start to go wrong (i.e. we have an unsafe act and/or unsafe condition) they act like dominos. A chain reaction takes place that cannot be stopped, resulting in accidents and injury.

Domino effect leading to incidents

Each year thousands of accidents are reported to the enforcement authorities. Some of the most common categories are:

- manual handling disorders
- slips, trips and falls
- being struck by a moving object.

Unfortunately 95% of all accidents are caused by human error and are therefore preventable.

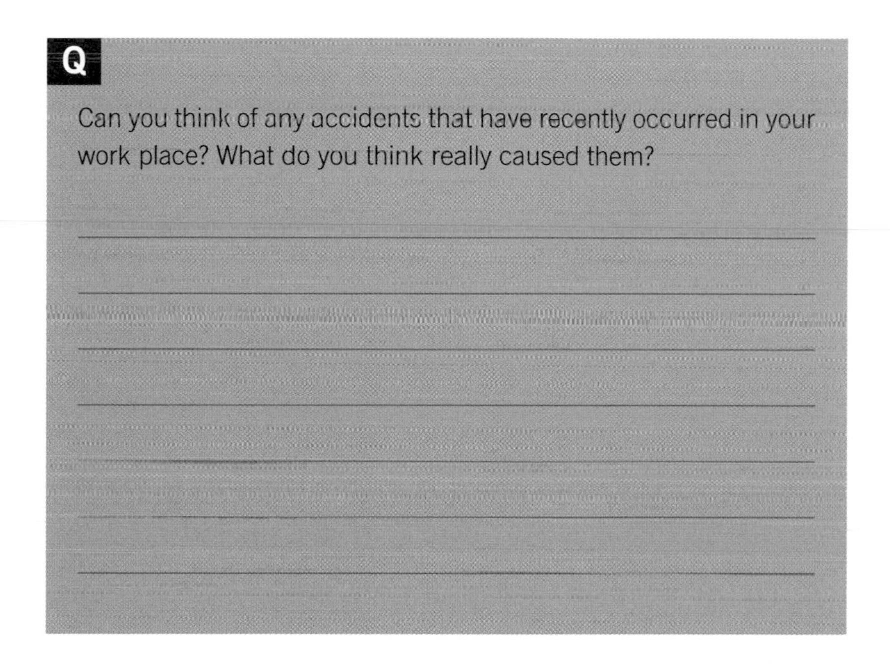

Once an accident, dangerous occurrence or near miss has occurred it is vital that it is reported for the following reasons:

1. Legislation

The Reporting of Injuries, Diseases and Dangerous Occurrences Regulations (RIDDOR) require employers to report the following to the local enforcing authority within specified time limits:

- fatal accidents
- fractures (not in the hand or foot)
- amputations
- loss of sight
- any injury resulting in immediate hospitalisation for more than 24 hrs
- specified dangerous occurrences such as a collapsing scaffold
- explosions
- specified workplace diseases that occur as a result of work activity.

The most common reason for employers to report an accident is when the injured person is off work for more than seven days as a result of the accident.

You as the employee need to notify your employer of all accidents that occur to you so that he/she can in turn fulfil his/her legal obligations.

2. Claim/claim defence

In order for an employee to make a claim from his/her employer or from the Department of Social Security, there must be documented evidence of the incident having occurred. The accident book is often used to ensure this process takes place.

The employer also needs to know that the incident occurred so that he/she has the opportunity to do something about it. When reporting an accident, as much information as possible must be recorded.

3. Hazard identification

By documenting and being aware of the incident, the employer is better able to ascertain the cause and to implement corrective action to prevent it happening again (this may include carrying out a risk assessment or amending an existing risk assessment).

4. Monitor trends

Having records of all accidents that have occurred in a place of work over a period of time allows your employer to:

- analyse the types of incident that have occurred;
- identify the staff involved and any training needs;
- ascertain the frequency of the incident;
- ascertain the seriousness of the incident;
- compare results with other similar businesses;
- compare results with previous years.

Research (carried out by Bird) has shown that for every accident resulting in a major injury there were approximately 10 resulting in minor injuries, 30 with property damage and 600 near misses.

WHY INVESTIGATE INCIDENTS AND ACCIDENTS?

By investing time in investigating incidents your employer can:

- identify the cause of the accident;
- improve safety standards within the workplace;
- prevent any recurrence by implementing a corrective action plan;
- defend fraudulent claims (employees have up to three years to claim for compensation);
- minimise the risk of prosecution by external enforcing authorities.

WHO SUFFERS FROM AN ACCIDENT?

The cost of having an accident can be counted in many ways, for example:

- loss of production
- plant/equipment damage
- physical harm
- psychological damage
- absenteeism
- loss of earnings
- permanent disability.

The real sufferers from the accident are the injured persons and their dependants. Always:

- **Think Safety** when carrying out your work
- **Work Safely** and you will be safe
- **Be Safe** and you will help avoid accidents involving yourself or others.

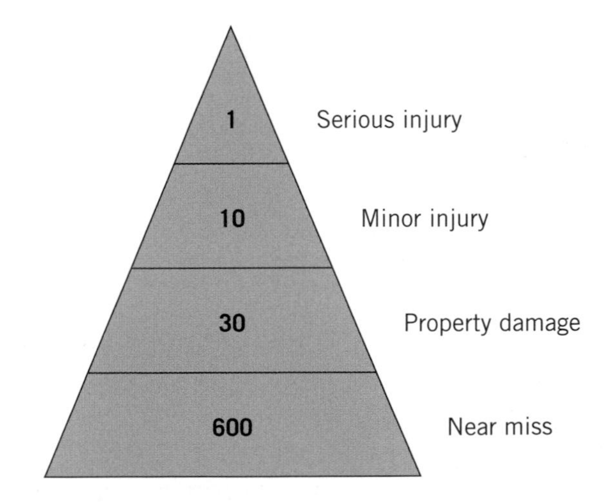

Bird's triangle

HAZARDS IN THE WORKPLACE

A damaged light fitting

A definition of a hazard is:

> "SOMETHING OR A SITUATION THAT HAS THE POTENTIAL TO CAUSE HARM."

In reality most things in the workplace or home have the potential to cause harm, for example:

- A single brick placed on a flat table in a room is relatively safe; however, the same brick falling from some high scaffolding could hit someone below and cause serious injury or even death.
- A strong acid in a properly labelled closed glass container in a laboratory cupboard would be relatively safe; however, the same acid left in a cup in a kitchen

could be accidentally drunk, causing serious burns to the mouth and tracts.

A simple change in situation or circumstances can change the nature of the hazard from safe to very unsafe. This is known as the risk associated with the hazard.

The definition of risk is:

"THE LIKELIHOOD OF THE HAZARD CAUSING HARM AND THE PROBABLE OUTCOME."

To help identification, hazards can be divided into five main categories within your workplace. These are:

1. **Physical:** These could be anything in the workplace you can feel or touch. Examples include:
   - doors
   - furniture
   - machinery
   - tools.
2. **Chemical:** These can be solid, liquid or gas. Examples include:
   - detergents/soaps
   - acids/alkalis
   - correction fluid
   - bleach
   - glue
   - oils
   - paints
   - solvents.
3. **Biological:** This could be bacterial material on work surfaces. Examples include:

- bacteria
- moulds.

4. **Ergonomic:** Relates to the ill health effects of poorly designed tasks and workstations leading to musculoskeletal problems including work-related upper limb disorders, whole body or hand/arm vibration syndrome.
5. **Psychological:** This includes mental health, workplace stress, violence at work, smoking and drug/alcohol abuse.

Hazards can be managed or controlled in a number of ways. The following list of control measures (in order of preference) can be targeted at any hazard. Ideally we should aim to:

- eliminate hazards all together by removing the hazard or changing the process;
- substitute hazardous materials or processes for less hazardous ones;
- keep people away from a hazard or the hazard away from the people, that is, prevent exposure;
- limit time spent in a hazardous area or carrying out a hazardous task to help limit the exposure;
- introduce local exhaust ventilation to help remove the hazard if it is gaseous in nature;
- improve general ventilation in the workplace to help dilute the hazard if it is gaseous in nature;
- improve housekeeping in the workplace to remove tripping and falling hazards;
- train employees to carry out the job correctly and safely; being fully aware of the hazards will reduce the likelihood of accidents and injury;
- ensure employees have adequate welfare facilities to help to reduce the hazards;

- where necessary and available, carry out health screening to help ensure other control measures are working and detect problems early;
- provide employees with personal protective equipment to help them reduce their exposure to the hazard providing that they are using the equipment in the right way.

Q

Could you, or does your employer, apply any of the above controls in your workplace?

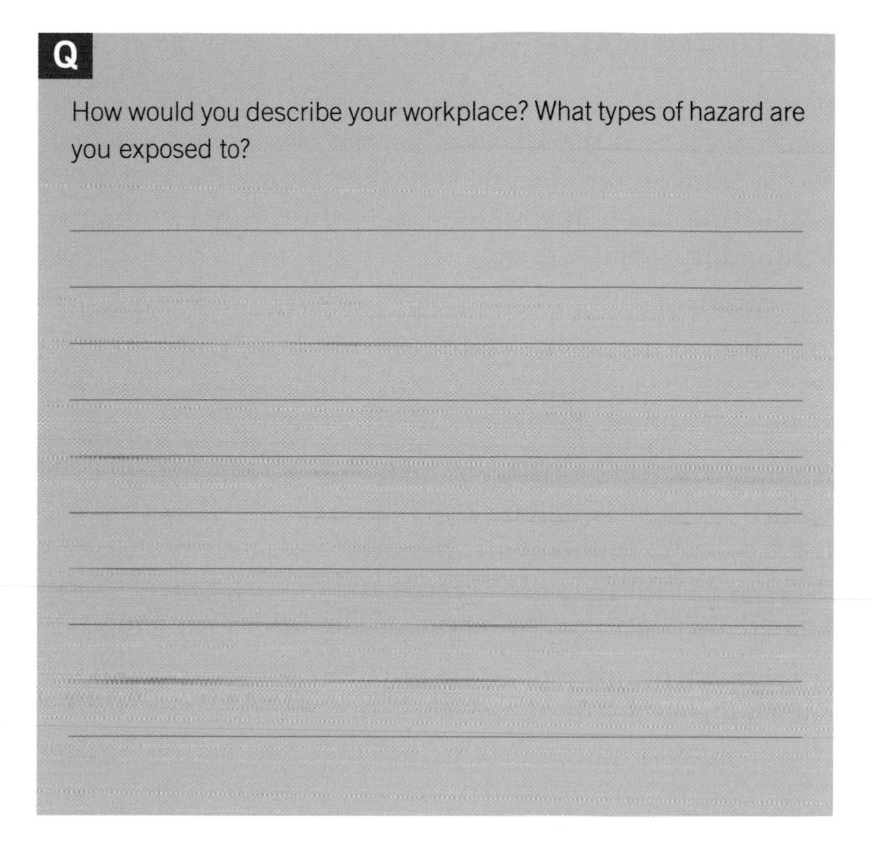

Q

How would you describe your workplace? What types of hazard are you exposed to?

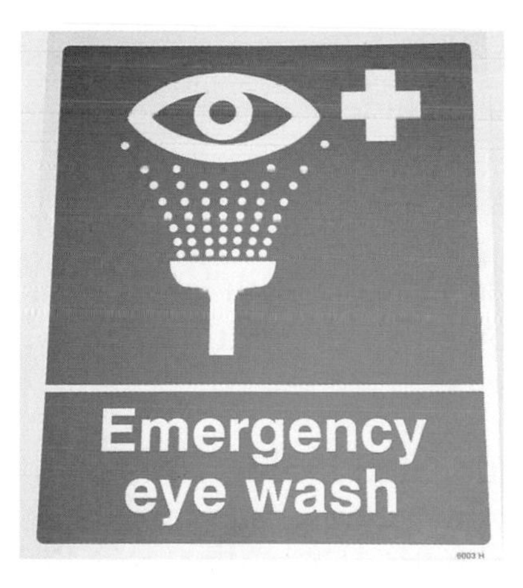

Eye wash signage

First aid signage

OCCUPATIONAL HEALTH

A typical workplace today is very different to the workplaces present in the 1970s when the Health and Safety at Work Act was introduced. There has been a marked move away from heavy manufacturing industry and mining to light industry and service.

As a result there has been a change in the types of injury and disease occurring. However, the legacy of some illnesses, such as deafness, asbestosis, lung cancer, that take years to manifest themselves is still present.

Although medical advances and technology have helped us reduce or even eliminate the risks associated with some hazards, it is unfortunate that we have in the last few decades introduced new hazards and risks. Of these work-related stress appears to be one of the fastest growing.

A worker's health will be directly affected by the type of environment he/she is working in. Occupational health surveillance is an important tool that can be used by an employer as part of his/her management strategy to help monitor the control measures in place and identify and diagnose potential health problems within the workforce early on so that they can be treated.

Examples of health surveillance include:

- audiometric tests for people working in noisy environments
- blood lead levels for people working with lead
- lung function tests for people working with asbestos
- eye sight tests for VDU users.

Unfortunately because occupational ill health issues often take many years to manifest themselves, they are often ignored by employers, and employees do not always realise they have a problem until it is too late.

EXERCISE 3

Self-assessment questions

1. What is the definition of risk?

....

....

2. Accidents and incidents occur as a result of what?

a) ....

b) ....

3. Give two examples of accidents that would be reportable to the enforcing authority.

a) ....

b) ....

4. List three different reasons for investigating accidents.

a) ....

b) ....

c) ....

5. What is the definition of a hazard?

....

....

....

....

6. List four of the five hazard categories.

a) ....

b) ....

c) ....

d) ....

7. Give two examples of health screening techniques.

a) ....

....

....

b) ....

....

....

You will find the answers on pages 94–5.

MODULE FOUR

Proactive health and safety

GOOD WORKING CONDITIONS ARE A BASIC HUMAN RIGHT AND A FUNDAMENTAL PART OF DECENT WORK WHILST WORKING SAFELY MUST BE SEEN AS A CONDITION OF EMPLOYMENT.

Welcome to module four. In this module you will learn about some of the practical steps you and your employer can take to help improve health and safety within the workplace and therefore reduce the risk of accidents and ill health. Remember, your employer can only succeed with your help, co-operation and commitment.

HEALTH AND SAFETY POLICY AND CLEAR RESPONSIBILITIES

Your employer must produce a written policy statement detailing his/her commitment to ensuring your and other people's health, safety and welfare. In addition he/she must set out clear responsibilities and accountabilities for key members of his/her staff to ensure that safety is really put into practice, and make arrangements to ensure that all safety procedures developed are implemented, monitored and reviewed as necessary.

AUDITS

Your employer may from time to time carry out audits to measure the effectiveness of the procedures and processes he/she has in place. These can help highlight deficiencies and gaps within his/her overall safety management system.

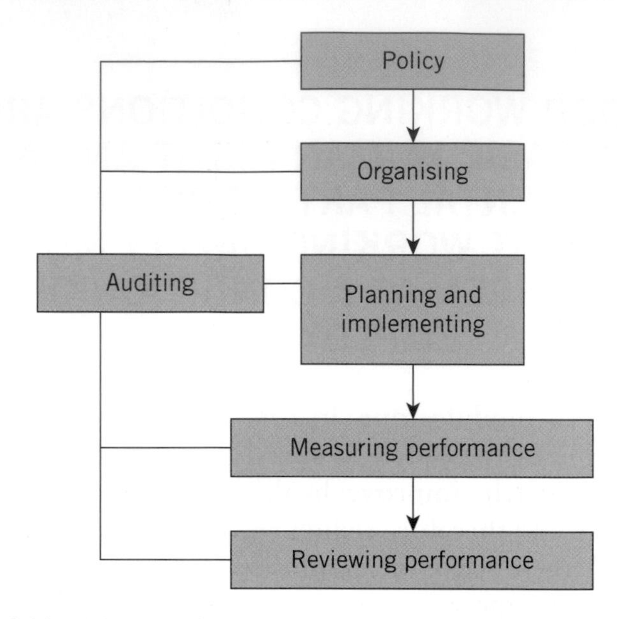

Successful health and safety management (HSG65)

INSPECTIONS

Health and safety inspections may be carried out using checklists and/or pro formas which measure conformance to the company's health and safety procedures in terms of documentation and the physical hazards present in the workplace. This allows your employer to "bench mark" and highlight the improvement or falling of standards ("performance tracking"). You may be asked to assist in the inspection process.

In reality you can carry out an inspection of your local place of work every day on an informal basis and by making your employer aware of any hazards could help to prevent an accident.

HAZARD/NEAR MISS REPORTING

As stated, your assistance is vital to improve safety standards: always report any hazards, near misses or dangerous situations you see or are aware of. In some instances you may even be able to resolve the issues yourself and help prevent an accident. Your employer may have a formal hazard/near miss reporting system in place. If so use it as it could help save a life or prevent an accident.

TRAINING

The purpose of training is to ensure that the employee is able to carry out the task/work activity to be performed consistently to an agreed standard and in a safe manner. All training, whatever the task, should cover any safety measures needed to ensure the task is performed safely, that is, the correct use of PPE, manual handling techniques, machine operating and so on.

Your employer may deliver the training on or off the job; it may be formal or informal, internal or external, in a group or on an individual basis. However when training is conducted, the purpose is to ensure that you learn. Your employer should document details of any training provided to prove that it took place.

Examples of when training is necessary include:

- induction training for all new employees to the business or new department/section;
- refresher training;
- specific training (e.g. first aid);
- training when new equipment is introduced.

A typical training room

SAFE SYSTEMS OF WORK/RISK ASSESSMENT

A safe system of work is a formal written procedure which results from a systematic examination of a task or work activity in order to identify all the hazards and associated risks involved in the task or activity (risk assessment). Once completed it should be used to provide you with instruction in the safe method of work and to ensure that the hazards are eliminated or the remaining risks are minimised.

The risk assessment identifies the key steps within the activity (how it is done); all the potential hazards are then identified (the "what if" chain of events) and assessed for risk. This is followed by an action plan of any preventative measures needed to eliminate or reduce the risks (the "what can be done to prevent accidents" part of the assessment).

Your employer must complete risk assessments for all tasks carried out, both routine and abnormal. Always ensure that you have been made aware of the relevant risks associated with the work you do and that you are aware of any particular precautions that must be taken.

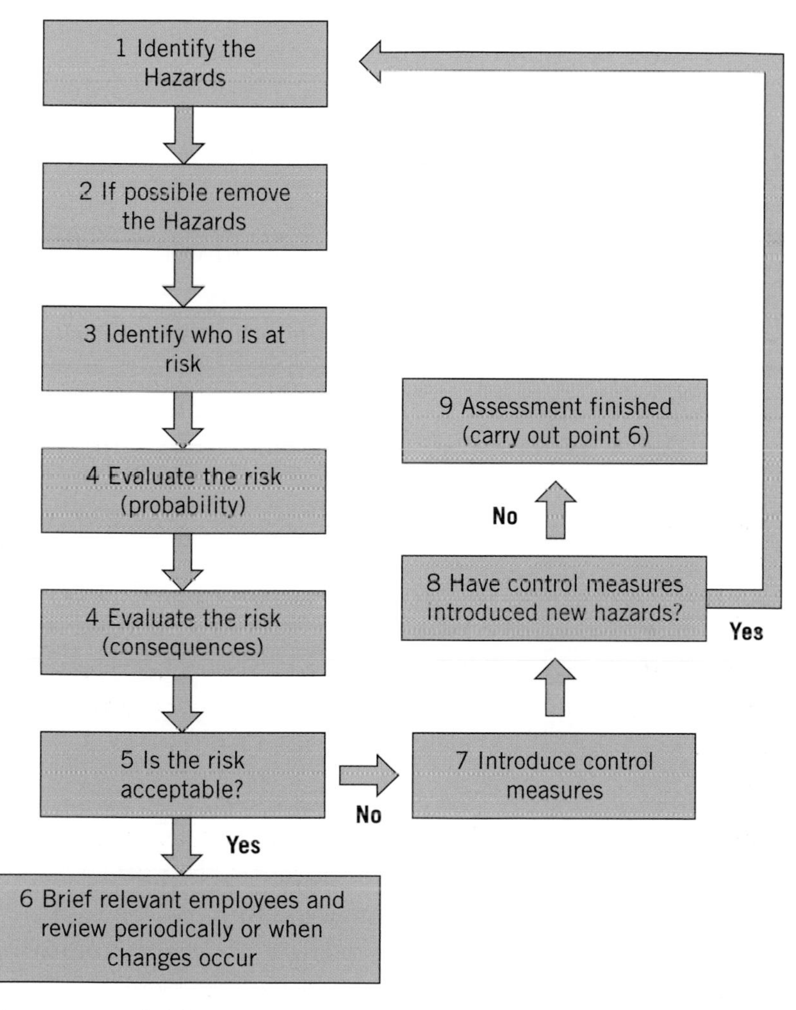

The process of risk assessment

PLANT/EQUIPMENT MAINTENANCE

Your employer has a duty to carry out or have carried out on his/her behalf specific tests and inspections to ensure that equipment/plant is safe to operate. Before using any plant or equipment ensure you have the necessary training/ knowledge and always carry out a quick visual inspection of the plant or equipment prior to use.

COMMUNICATION

Your employer has a legal duty to consult with you on matters relating to health and safety directly or through nominated representatives (these may be unionised workplace representatives or non-unionised workplace representatives). It is in his/her interest and yours to ensure that you are fully aware of all hazards and risks present within the workplace.

Remember, communication is a two-way process. It may be achieved in many ways, such as:

- formal meetings
- informal meetings
- newsletters
- tool box talks
- training sessions
- notice boards
- e-mail.

Ensure you know who your representatives are and how you can communicate matters relating to health and safety to your employer.

PLANNED MAINTENANCE

By developing and implementing a planned maintenance programme your employer cannot only comply with legislation but also help ensure that work equipment is likely to be safe when properly used.

COOPERATION

The management of health and safety is a two-way process requiring the commitment and co-operation of both employer and employee.

You can help your employer improve safety standards and reduce accidents by the positive actions you take.

EXERCISE 4

Self-assessment questions

1. Give two examples of how inspections can help manage health and safety.

a) ....

....

b) ....

....

2. Is there a legal requirement to carry out risk assessments?

a) Yes
b) No

3. List three different ways of delivering training.

a) ....

b) ....

c) ....

4. List three ways the employer can communicate with his/her employees

a) ....

b) ....

c) ....

5. Why is it important to report hazards and near misses?

....

....

....

6. Does your employer have to produce a safety policy statement?

a) Yes
b) No

7. When is it necessary to provide training?

....

....

....

You will find the answers on pages 95–6.

Answers to questions

MODULE 1 – Health and safety law and enforcement

1. What is the maximum fine you can receive at a magistrates' court?

c) £20,000

2. What are the two important types of law?

c) Civil & Criminal

3. Can an enforcement officer take equipment away from your workplace?

a) Yes

4. List four reasons for employers to address health and safety.

Moral, Legal/law, Financial/cost, Business need

5. Give two reasons why employees should take health and safety seriously.

Legal requirement, moral, financial benefits, business needs

6. Who enforces health and safety law in the UK?

c) Environmental Health Officers and Factory Inspectors

7. Health and safety law places a duty on:

c) Employers, employees and the self-employed

MODULE 2 – Health and safety regulations

1. What are the four colours used for safety signs?

Blue, Yellow, Red and Green

2. What are the two key action levels under the Noise at Work Regulations?

b) 80 & 85dB(A)

3. Can exposure to one very loud noise cause deafness?

a) Yes

4. List four different health and safety control measures that can be applied to reduce risk.

Elimination, substitution, limit exposure, local exhaust ventilation, health surveillance, general ventilation, housekeeping

5. Which muscles should be used when lifting a load?

b) Leg muscles

6. What is the purpose of first aid?

To preserve life and prevent the deterioration of the casualty

7. List two things that must be assessed when carrying out a VDU assessment.

The screen, keyboard, chair, desk, environment

MODULE 3 – Accidents and incidents

1. What is the definition of risk?

The likelihood of a hazard causing harm and the probable consequences

2. Accidents and incidents occur as a result of what?

Unsafe acts, unsafe conditions

3. Give two examples of accidents that would be reportable to the enforcing authority.

   Fractures, amputations, death, over three-day absence due to injury, 24 hour hospitalisation, unconsciousness, loss of sight

4. List three different reasons for investigating accidents.

   Identify cause, prevent recurrence, claim defence, insurer stipulation

5. What is the definition of a hazard?

   Something that has the potential to cause harm

6. List four of the five hazard categories.

   Physical, chemical, biological, ergonomic, psychological

7. Give two examples of health screening techniques.

   Audiometry, lung function testing, chest x-rays, blood sampling

MODULE 4 – Proactive health and safety

1. Give two examples of how inspections can help manage health and safety.

   Identify hazards, prevent accidents, demonstrate commitment

2. Is there a legal requirement to carry out risk assessments?

   a) Yes

3. List three different ways of delivering training.

   One to one, on the job, off the job, group training, distance learning, computer-based training, verbal instructions

4. List three ways the employer can communicate with his/her employees.

Tool-box talks, notice boards, newsletters, verbal instructions, meetings, videos

5. Why is it important to report hazards and near misses?

To prevent them from becoming accidents in the future

6. Does your employer have to produce a safety policy statement?

a) Yes

7. When is it necessary to provide training?

Induction, change of job, promotion, refresher, new equipment or change in process

Glossary/definitions

Accident An unplanned and/or uncontrolled event which has led to injury, damage or other loss to the business

COSHH Control of Substances Hazardous to Health regulations

dB Decibel: the measure of sound

DSE Health and Safety (Display Screen Equipment) Regulations

EHO Environmental Health Officer

EMAS Employment Medical Advisory committee

EPA Environmental Protection Act

FFI Fee for Intervention

Hazard Something that has the potential to cause harm

H&S Health and Safety

HSC Health and Safety Commission

HSE Health and Safety Executive

HSWA Health and Safety at Work Act

MHOR Manual Handling Operations Regulations

MHSWR Management of Health and Safety at Work Regulations

MSDS Material Safety Data Sheet

PAT Portable Appliance Testing

PPE Personal Protective Equipment

PUWER Provision and Use of Work Equipment Regulations

RIDDOR Reporting of Injuries, Diseases and Dangerous Occurrences Regulations

Risk The probability and consequences of a hazard occurring

RPE Respiratory Protective Equipment

Six Pack Term given to a group of six regulations introduced in 1993: **MHSWR, MHOR, PPE, DSE, PUWER, WHSWR**

SHE Safety, Health and Environment

UK United Kingdom

VDU Visual Display Unit

WEL Workplace Exposure Limit

WHSWR The Workplace (Health, Safety and Welfare) Regulations

WRULDs Work Related Upper Limb Disorders

Examination

Once you have completed the guide you can choose to complete the examination.

Complete the following exam to test your health and safety workplace knowledge. All candidates will be informed whether they have passed or failed via email, and candidates achieving a pass mark of 75% or above will receive a certificate.

To find out details of where to send your completed exam, or to request an electronic version, please visit www.routledge.com/9780415835442.

You have forty minutes – good luck

Your FULL Name: (PRINT)		Employee Reference:	Employer Name:
Date of Birth:		Your Email Address:	
Date of Examination:			

Please indicate your answer(s) by putting a tick in the appropriate box(es)

1. What are the TWO types of law relating to health and safety?

   a) Criminal ❐
   b) Statute ❐
   c) Civil ❐
   d) Common ❐

2. Under the HASAW Act, what are the TWO enforcement notices that are issued?

   a) Confiscation order ❐
   b) Prohibition ❐
   c) Renewal ❐
   d) Improvement ❐

3. What is the maximum fine that can be imposed for a health and safety offence in a **magistrates' court**?

    a) £5,000 ❐
    b) £10,000 ❐
    c) £20,000 ❐
    d) £40,000 ❐

4. What is meant by the term **reasonably practicable**?

    a) The balance between risk and health ❐
    b) The balance between risk and cost ❐
    c) The balance between cost and health ❐
    d) The balance between cost and prosecution ❐

5. Which one of the following **is not** a duty of an employer under the Health and Safety at Work Act?

    a) Providing information, instruction and training for employees ❐
    b) Providing a safe place of work ❐
    c) Providing safe systems of work ❐
    d) Consulting with employees ❐

6. Which one of the following **is not** a duty of an employee under the Health and Safety at Work Act?

    a) To buy personal protective equipment for themselves ❐
    b) Not to interfere or misuse anything provided for Health and Safety ❐
    c) To care for the health and safety of themselves and other people ❐
    d) to co-operate with his/her employer ❐

7. Who <u>enforces</u> Health and Safety legislation in the UK?

   a) Factory Inspectors and firemen ❒
   b) Policemen and Firemen ❒
   c) Factory Inspectors and Environmental Health Officers ❒
   d) Environmental Health Officers and firemen ❒

8. Which of the following <u>is not</u> a power of a Factory Inspector?

   a) To visit any workplace ❒
   b) To arrest an employee ❒
   c) To carry out investigations and examinations ❒
   d) To take statements ❒

9. What <u>message</u> does a Red safety sign give?

   a) Warning ❒
   b) Mandatory ❒
   c) Prohibition ❒
   d) Safe condition ❒

10. What are the <u>two key</u> action levels under the Noise at work regulations?

    a) 80dB and 85dB ❒
    b) 85dB and 90dB ❒
    c) 80dB and 90dB ❒
    d) 75dB and 80dB ❒

11. Which <u>muscles</u> should be used when lifting a load?

    a) The arm muscles ❒
    b) The leg muscles ❒
    c) The back muscles ❒
    d) The stomach muscles ❒

12. What are the **main purposes** of first aid?

a) To cure the casualty's illness ❒
b) To preserve life and prevent the deterioration of the casualty ❒
c) To injure the casualty ❒
d) To perform miracles ❒

13. Which **four factors** must be taken into account under the Manual Handling Operations Regulations?

a) The task, load, working environment and individual capability ❒
b) The weight, size, temperature and dimensions ❒
c) The temperature, room space, individual capability and colour ❒
d) The task, load, shape and time ❒

14. What are the **dangers** associated with electricity?

a) Death ❒
b) Electrical burns and shock ❒
c) Fires ❒
d) All of the above ❒

15. What is the **most effective** way of controlling a hazard and its associated risk?

a) Prevent exposure to it ❒
b) Substitute it ❒
c) Eliminate it ❒
d) Provide personal protective equipment ❒

16. What is the minimum acceptable temperature in an office?

    a) 25° C ❐
    b) 11° C ❐
    c) 16° C ❐
    d) 21° C ❐

17. Which one of the following injuries is not reportable to the enforcing authorities?

    a) Death ❐
    b) Fractured arm ❐
    c) Amputated leg ❐
    d) Fractured finger ❐

18. Which one of the following items must be assessed when carrying out a display screen equipment assessment?

    a) The operator's clothing ❐
    b) The keyboard ❐
    c) The operator's shoes size ❐
    d) The number of plants in the room ❐

19. What is the definition of risk?

    a) Something that will cause harm ❐
    b) Something that has the potential to cause harm ❐
    c) The probability of harm being caused and the likely consequences ❐
    d) Being struck by a falling brick ❐

20. What is the definition of a hazard?

    a) Something that will cause harm ❐
    b) Something that has the potential to cause harm ❐
    c) The probability of harm being caused and the likely consequences ❐
    d) Being struck by a falling brick ❐

21. Which one of the following is not a typical example of occupational health monitoring?

    a) Audiometric testing ❐
    b) Lung function testing ❐
    c) Blood lead level testing ❐
    d) Employee weight checks ❐

22. Which one of the following is not a primary reason for investigating accidents?

    a) Defending a claim for compensation ❐
    b) Identifying the cause of the accident ❐
    c) Helping to prevent a reoccurrence of the accident ❐
    b) Apportioning blame to employees ❐

23. How long does an employee have to make a compensation claim following an accident at work?

    a) One year ❐
    b) Three years ❐
    c) Five years ❐
    d) Ten years ❐

24. What **three** elements are required to produce a fire?

   a) Heat, light and smoke ❐
   b) Heat, fuel and oxygen ❐
   c) Fuel, nitrogen and heat ❐
   d) Oxygen, light and air ❐

25. On which type of fire would you **not** use a water-based fire extinguisher?

   a) Paper-based fire ❐
   b) Wooden-furniture-based fire ❐
   c) Electrical-appliance-based fire ❐
   d) Textiles-based fire ❐

26. Which one of the following is **not** a proactive safety measure?

   a) Carrying out regular audits ❐
   b) Carrying out regular training ❐
   c) Carrying out regular workplace inspections ❐
   d) Investigating an accident ❐

27. When is it **not** appropriate for an employer to provide health and safety training for his/her staff?

   a) When starting a new job ❐
   b) After an accident ❐
   c) When a new machine has been introduced to the workplace ❐
   d) When leaving the company ❐

28. Which of the following safety measures is an example of reactive monitoring?

a) Carrying out regular audits ❒
b) Carrying out regular training ❒
c) Carrying out regular workplace inspections ❒
d) The analysis of accident data ❒

29. According to Bird's triangle how many near misses equate to one serious injury?

a) 500 ❒
b) 600 ❒
c) 450 ❒
d) 700 ❒

30. Which of the following is not an initiative to help the environment?

a) Turning off all unnecessary electrical machinery at work ❒
b) Recycling waste paper and cardboard ❒
c) Reporting all dripping taps at work ❒
d) Disposing of waste chemicals into the local river ❒

Well done, you have now finished the test. Please check to ensure that you have answered all 30 questions (remember, some questions require two answers).